Charles Barney Cory

A List of the Birds of the West Indies

Charles Barney Cory

A List of the Birds of the West Indies

ISBN/EAN: 9783744717762

Printed in Europe, USA, Canada, Australia, Japan

Cover: Foto ©berggeist007 / pixelio.de

More available books at **www.hansebooks.com**

A

LIST OF THE BIRDS

OF THE

WEST INDIES,

INCLUDING THE

BAHAMA ISLANDS AND THE GREATER AND LESSER ANTILLES, EXCEPTING
THE ISLANDS OF TOBAGO AND TRINIDAD.

BY

CHARLES B. CORY.

REVISED EDITION.

ESTES & LAURIAT,
BOSTON, U. S. A.
1886.

INDEX TO FAMILIES.

FAMILY

TURDIDÆ.

Genus TURDUS.

MUSTELINUS. Gmel. . . . . Cuba.
FUSCESCENS. Steph. . . . . Cuba.
SWAINSONI. Cab. . . . . Cuba.
ALICIÆ. Baird. . . . . Cuba and San Domingo.

Genus MERULA.

MIGRATORIA. (Linn.) . . . . Cuba.
AURANTIA. (Gmel.) . . . . Jamaica.
JAMAICENSIS. (Gmel.) . . . . Jamaica.
GYMNOPHTHALMA. (Caban.) . . . . Grenada.
NIGRIROSTRIS. (Lawr.) . . . . St. Vincent.

Genus MIMOCICHLA.

RUBRIPES. (Temm,) . . . . Cuba.
SCHISTACEA. Baird . . . . Cuba.
PLUMBEA. (Linn.) . . . . Bahamas.
ARDESIACA. (Vieill.) . . . . . San Domingo and Porto Rico.

Genus CICHLHERMINIA.

HERMINIERI. (Lafr.) . . . Martinique and Guadeloupe.
SANCTÆ-LUCIÆ. (Scl.) . . . . St. Lucia.
DOMINICENSIS. (Lawr.) . . . . Dominica.

Genus SIALIA.

SIALIS. (Linn.) . . . . . Cuba.

Genus MYIADESTES.

SIBILANS. Lawr. . . . . St. Vincent.
GENIBARBIS. Sw. . . . . Martinique.
SANCTÆ-LUCIÆ. Stejn. . . . . St. Lucia.
DOMINICANUS. Stejn. . . . . Dominica.
MONTANUS. Cory . . . . Haiti.
SOLITARIUS. Baird . . . . Jamaica
ELIZABETH. (Lemb.) . . . . Cuba.
?ARMILLATUS. (Vieill.) . . . . West Indies ?

6

FAMILY

MIMIDÆ.

Genus MARGAROPS.

FUSCATUS. (Vieill.)   Porto Rico, Inagua, St. Thomas, and San Domingo?
DENSIROSTRIS. (Vieill.)
    Antigua, Dominica, Martinique, Montserrat, St. Lucia, Guadeloupe.
MONTANUS. (Vieill.)
    Martinique, St. Vincent, Dominica, St. Lucia, Guadeloupe.

Genus RAMPHOCINCLUS.

BRACHYURUS. (Vieill.) .   . .   St. Lucia, Martinique.

Genus CINCLOCERTHIA.

RUFICAUDA. (Gould) . . .   . Guadeloupe, Dominica.
MACRORHYNCHA. (Scl.)   .   . . . . St. Lucia.
GUTTURALIS. (Lafr.) .   .   . . . Martinique.

Genus GALEOSCOPTES.

CAROLINENSIS. (Linn.) .   . . . .   . Bahamas, Cuba.

Genus MIMUS.

POLYGLOTTUS. (Linn.) .   . . . Cuba.
ORPHEUS. (Linn.) .   .   . . . . Jamaica, Porto Rico.
ELEGANS. Sharpe . .   .   . . Inagua, Bahamas.
DOMINICUS. (Linn.)   .   . . . . San Domingo.
GILVUS. (Vieill.) . .   .   St. Lucia, St. Vincent, Grenada.
HILLII. March . . .   . Jamaica.
GUNDLACHI. Cab. . .   .   . . . Cuba and Bahamas.

FAMILY

SYLVIIDÆ.

Genus POLIOPTILA.

CÆRULEA. (Linn.) .   . . . . .   Cuba and Bahamas.
LEMBEYI. Gundl. .   . . . Cuba.

FAMILY
TROGLODYTIDÆ.

Genus THRYOTHORUS.

MARTINICENSIS. Scl.	Martinique.
RUFESCENS. Lawr. .	Dominica and Guadeloupe.
MUSICUS. Lawr. .	St. Vincent.
GRENADENSIS. Lawr.	. Grenada.
MESOLEUCUS. Scl. .	St. Lucia.

FAMILY
MNIOTILTIDÆ.

Genus MNIOTILTA.

VARIA. (Linn.) .   Bahamas and Antilles.

Genus PARULA.

AMERICANA. (Linn.)   Bahamas and Greater Antilles.

Genus PROTONOTARIA.

CITREA. (Bodd.)   . Cuba.

Genus HELMITHERUS.

VERMIVORUS. (Gmel.) .   Cuba and Jamaica.
SWAINSONI. (Aud.) .   Cuba and Jamaica.

Genus HELMINTHOPHAGA.

CHRYSOPTERA. (Linn.) .   . Cuba.
BACHMANI. (Aud.) .   . Cuba.
PEREGRINA. (Wils.) .   . Bahamas, Cuba.

Genus PERISSOGLOSSA.

TIGRINA. (Gmel.) .   . Bahamas, Greater Antilles, St. Croix.

Genus DENDRŒCA.

ÆSTIVA. (Gm.) . . . . Cuba and Bahamas?
PETECHIA. (Linn.) . . . . . . Bahamas and Antilles.
  petechia gundlachi. (*Baird.*) *Ridgw.* . . . Bahamas and Cuba.
  petechia ruficapilla. (*Gmel.*) *Ridgw.*
    . Porto Rico, Barbuda, St. Thomas, and Antigua.
  petechia melanoptera. *Lawr.* . . . . . Guadeloupe and Dominica.
CAPITALIS. Lawr. . . . . Barbadoes.
RUFIGULA. Baird . . . . Martinique.
EOA. Gosse . . . . Jamaica.
CÆRULESCENS. (Gmel.) . . . . . Bahamas, Greater Antilles.
CORONATA. (Linn.) . . . . . . Bahamas, Greater Antilles.
MACULOSA. (Gmel.) . . . . . Bahamas, Cuba, and San Domingo.
CÆRULEA. Wils. . . . . Cuba.
PENNSYLVANICA. (Linn.) . . . . Bahamas.
STRIATA. (Forst.) . . . Porto Rico, Cuba, Jamaica, and Bahamas.
PHARETRA. (Gosse) . . . . Jamaica.
BLACKBURNIÆ. (Gmel.) . . . . Bahamas.
DOMINICA. (Linn.) . . Porto Rico, Cuba, San Domingo, and Bahamas.
ADELAIDÆ. Baird . . . . Porto Rico and Antilles.
  adelaidæ delicata. *Ridgw.* . . . . St. Lucia.
VIRENS. (Gmel.) . . . . . . Cuba, Jamaica, and Dominica.
KIRTLANDI. Baird . . . . New Providence and Andros, Bahamas.
PITYOPHILA. (Gundl.) . . . . Cuba.
PINUS. (Wils.) . . . . Bahamas and San Domingo.
DISCOLOR. (Vieill.) . . . . Antilles.
PALMARUM. (Gmel). . . . . . Bahamas and Greater Antilles.
PLUMBEA. Lawr. . . . . . . Guadeloupe and Dominica.

Genus LEUCOPEZA.

SEMPERI. Scl. . . . . St. Lucia.

Genus CATHAROPEZA.

BISHOPI. (Lawr.) . . . . St. Vincent.

Genus SEIURUS.

AUROCAPILLUS. (Linn.) . . . . . . Bahamas and Antilles.
NOVEBORACENSIS. (Gmel.) . . . . . Bahamas and Antilles.
MOTACILLA. (Vieill.) . . . San Domingo, Cuba, and Jamaica.

Genus OPORORNIS.

FORMOSUS. (Wils.) . . . . Cuba.

Genus GEOTHLYPIS.

TRICHAS. (Linn.)	Bahamas and Antilles.
ROSTRATUS. Bryant	New Providence, Bahamas.

Genus MICROLIGEA.

PALUSTRIS. Cory	San Domingo.

Genus TERETISTRIS.

FERNANDINÆ. Lemb.	. Cuba.
FORNSI. Gundl.	. Cuba·

Genus SYLVANIA.

MITRATUS. (Gmel.)	Cuba and Jamaica.

Genus SETOPHAGA.

RUTICILLA. (Linn.)	Bahamas and Antilles.

FAMILY

CŒREBIDÆ.

Genus CERTHIOLA.

BAHAMENSIS. Reich.	Bahamas.
FLAVEOLA. (Linn.)	Jamaica.
PORTORICENSIS. (Bryant)	Porto Rico and St. Thomas.
? SANCTI-THOMÆ. Ridgw.	. St. Thomas.
BARTHOLEMICA. (Sparrm.)	St. Bartholomew.
NEWTONI. Baird	. St. Croix.
BANANIVORA. (Gmel.)	San Domingo.
DOMINICANA. (Taylor) Barbuda, Antigua, Dominica, Guadeloupe, and Nevis.
? SUNDEVALLI. Ridgw.	Guadeloupe.
MARTINICANA. Reich.	. St. Lucia and Martinique.
? FINSCHI. Ridgw.	. Dominica or Martinique.?
BARBADENSIS. Baird.	. Barbadoes.
SACCHARINA. Lawr.	St. Vincent.
ATRATA. Lawr.	. Grenada and St. Vincent.

Genus CŒREBA.

CYANEA. (Linn.)	. . Cuba.?

Genus GLOSSIPTILA.

RUFICOLLIS. (Gmel.)	Jamaica.

Genus CHLOROPHANES.

ATRICAPILLA. (Vieill.) (Scl. et Salv.)	. Cuba.?

FAMILY ·
HIRUNDINIDÆ.

Genus PROGNE.

DOMINICENSIS. (Gmel.) . . . . GREATER ANTILLES.
SUBIS. (Linn.) . . . . CUBA.

Genus PETROCHELIDON.

FULVA. (Vieill.) . . . . ANTILLES.

Genus HIRUNDO.

BICOLOR. Vieill. . . . . CUBA and BAHAMAS.
CYANEOVIRIDIS. Bryant . . . . BAHAMAS.
EUCHRYSEA. Gosse . . . . JAMAICA.
SCLATERI. Cory . . . . SAN DOMINGO.
ERYTHROGASTRA. (Bodd.) GREATER ANTILLES, ST. CROIX, and GUADELOUPE.

Genus COTYLE.

RIPARIA. (Linn.) . . . . . CUBA, JAMAICA, and ANTILLES.

FAMILY
VIREONIDÆ.

Genus VIREO.

MODESTUS. Scl. . . . . JAMAICA.
LATIMERI. Baird . . . . PORTO RICO.
CRASSIROSTRIS. (Bryant) . . . . BAHAMAS.
GUNDLACHI. Lemb. . . . . CUBA.
NOVEBORACENSIS. (Gmel.) . . . . CUBA.
FLAVIFRONS. Vieill. . . . . CUBA.
CALIDRIS. (Linn.) . . . . ANTILLES.

calidris barbatulus. (Cab.) Bd. . . . . . . BAHAMAS and CUBA.

OLIVACEA. (Linn.) . . . . . . CUBA and PORTO RICO.
SOLITARIUS. (Wils.). . . . . CUBA.

Genus LALETES.

OSBURNI. Scl. . . . . JAMAICA.

FAMILY

AMPELIDÆ.

Genus DULUS.

DOMINICUS. (Linn.) . . SAN DOMINGO.
NUCHALIS. Sw. . . . . . W. I.?

Genus AMPELIS.

CEDRORUM. Vieill. . . . . . . Cuba and Jamaica.

FAMILY

TANAGRIDÆ.

Genus EUPHONIA.

MUSICA. (Gmel.) . . . . San Domingo.
FLAVIFRONS. (Sparrm.) St. Bartholomew, Martinique,
    Guadeloupe, Dominica, St. Vincent, and Grenada.
JAMAICA. (Linn.) . . . . Jamaica.
SCLATERI. Bp. . . . . Porto Rico.

Genus CALLISTE.

CUCULLATA. (Swain.) . . . . St. Vincent.

Genus SPINDALIS.

ZENA. (Linn.) . . . . Bahamas.
PRETRII. (Less.) . . . . Cuba.
MULTICOLOR. (Vieill.) . . . . San Domingo.
PORTORICENSIS. (Bryant) . . . . . . Porto Rico.
NIGRICEPHALA. (Jameson) . . . . . . Jamaica.

Genus PYRANGA.

ÆSTIVA. (Gmel.) . . . . . . Bahamas and Cuba.
RUBRA. (Linn.) . . . . . Jamaica, Cuba and Barbadoes.

Genus NEOSPINGUS.

SPECULIFERUS. Lawr. . . . . Porto Rico.

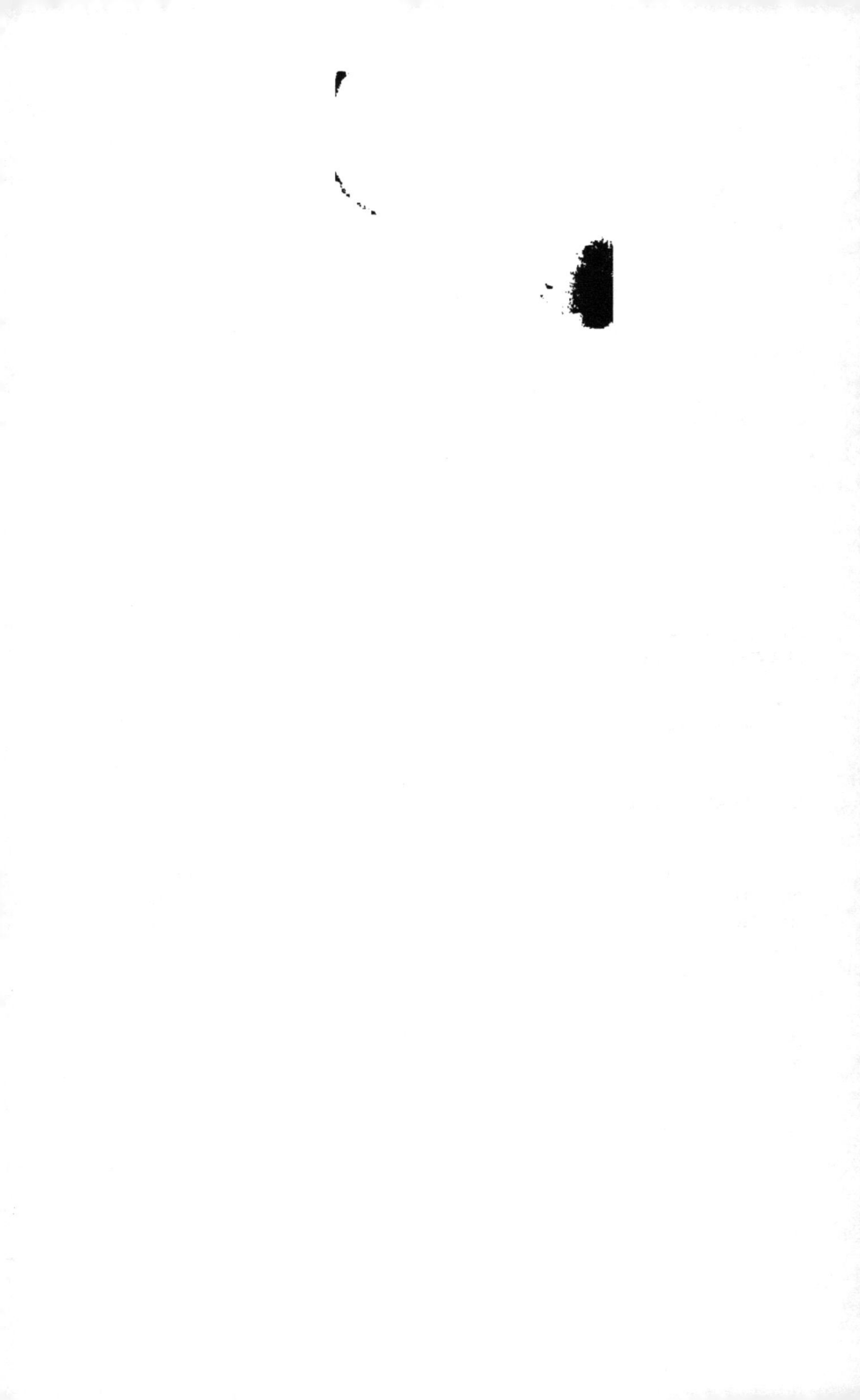

GENUS PHŒNICOPHILUS.

PALMARUM. (LINN.).	. SAN DOMINGO.
DOMINICENSIS. CORY	. SAN DOMINGO.

GENUS CALYPTOPHILUS.

FRUGIVORUS. CORY . . . .   SAN DOMINGO.

GENUS SALTATOR.

GUADALOUPENSIS. LAFR. . . . . . . GUADELOUPE and MARTINIQUE.

FAMILY

FRINGILLIDÆ.

GENUS GUIRACA.

CÆRULEA. (LINN.) . . . . CUBA.

GENUS HABIA.

LUDOVICIANA. (LINN.) . . . . . .   CUBA and JAMAICA.

GENUS LOXIGILLA.

VIOLACEA. (LINN.) . . . . . . BAHAMAS, JAMAICA, and SAN DOMINGO.
NOCTIS. (LINN.) . . . . LESSER ANTILLES.
noctis sclateri. Allen. . . . . . . ST. LUCIA.
ANOXANTHA. (GOSSE) . . . . JAMAICA.
PORTORICENSIS. DAUD. . . . . PORTO RICO.
portoricensis grandis. Lawr. . . . . . ST. CHRISTOPHER.

GENUS MELOPYRRHA.

NIGRA. (LINN.) . . . . CUBA.

GENUS LOXIMITRIS.

DOMINICENSIS. BRYANT . . . . SAN DOMINGO.

GENUS CHRYSOMITRIS.

CUCCULLATA. (SWAINS.) (Introduced) . . . . PORTO RICO and CUBA.

GENUS EUETHEIA.

OLIVACEA. (GMEL) . . . . CUBA, JAMAICA, HAITI, and PORTO RICO.
CANORA. (GMEL.) . . . . CUBA.
BICOLOR. (LINN.) . . . . BAHAMAS and ANTILLES.
ADOXA? . . . . JAMAICA.

GENUS PASSERINA.

CIRIS. (LINN.) . . . . CUBA and BAHAMAS.
CYANEA. (LINN.) . . . . CUBA and BAHAMAS.

Genus PASSER.

DOMESTICUS. (Linn.) (Introduced) . BAHAMAS and ANTILLES.

Genus PASSERCULUS.

SAVANNA. (Wils.) . . . . BAHAMAS and CUBA.

Genus SPIZELLA.

SOCIALIS. (Wils.) . . . . . . CUBA.

Genus COTURNICULUS.

PASSERINUS. (Wils.) . JAMAICA, CUBA, PORTO RICO, BAHAMAS.

Genus SYCALIS.

FLAVEOLA. (Linn.) . . . . . . JAMAICA.

Genus HABROPYGA.

MELPODA. (Vieill.) (Introduced) . . . . PORTO RICO.

FAMILY
PLOCEIDÆ.

Genus SPERMESTES.

CUCULLATUS. (Swain.) (Introduced) . . . . PORTO RICO.

FAMILY
ICTERIDÆ.

Genus ICTERUS.

BONANA. (Linn.) . . . . MARTINIQUE.
HYPOMELAS. Bp. . . . . CUBA.
DOMINICENSIS. (Linn.) . . . . HAITI.
PORTORICENSIS. Bryant . . . . PORTO RICO.
LAUDABILIS. Scl. . . . . ST. LUCIA.
CUCULLATUS. Sw. . . . . CUBA.
LEUCOPTERYX. (Wagl.) . . . . JAMAICA.
SPURIUS. Linn. . . . . CUBA.
OBERI. Lawr. . . . . MONTSERRAT.
VULGARIS. Daud. (Introduced) . . . . PORTO RICO and JAMAICA.
GALBULA. (Linn.) . . . . CUBA.

GENUS DOLICHONYX.

ORYZIVORUS. (LINN.) . . . . . . CUBA and JAMAICA.

GENUS AGELÆUS.

HUMERALIS. (VIG.) . . . . CUBA.
XANTHOMUS. (SCL.) . . . . PORTO RICO.
PHŒNICEUS. .(LINN.) . . . . BAHAMAS.
ASSIMILIS. GUNDL. . . . . CUBA.

GENUS XANTHOCEPHALUS.

ICTEROCEPHALUS . . . . CUBA.

GENUS STURNELLA.

HIPPOCREPIS. WAGL. . . . . CUBA.

GENUS NESOPSAR.

NIGERRIMUS. (OSBURN) . . . JAMAICA.

GENUS QUISCALUS.

FORTIROSTRIS. LAWR. . . . . BARBADOES.
INFLEXIROSTRIS. SW. . . . . . ST. LUCIA and MARTINIQUE.
BRACHYPTERUS. CASS. . . . . PORTO RICO.
CRASSIROSTRIS. SW. . . . . JAMAICA.
LUMINOSUS. LAWR. . . . . GRENADA.
GUADELOUPENSIS. LAWR. . . . . . . GUADELOUPE.
GUNDLACHI. CASS. . . . . CUBA.
NIGER. (BODD.) . . . . SAN DOMINGO.
ATROVIOLACEUS. D'ORB. . . . . CUBA.

FAMILY

CORVIDÆ.

GENUS CORVUS.

LEUCOGNAPHALUS. DAUD. . . . . . HAITI and PORTO RICO.
JAMAICENSIS. GMEL. . . . . JAMAICA.
SOLITARIUS. WÜRT. . . . . SAN DOMINGO.
NASICUS. TEMM. . . . . CUBA.
OSSIFRAGUS. WILS. . . . . CUBA.
MINUTUS. GUNDL. . . . . CUBA.

FAMILY

TYRANNIDÆ.

GENUS ELAINEA.

MARTINICA. (LINN.).	. LESSER ANTILLES.
FALLAX. SCL.	. . JAMAICA.
COTTA. GOSSE	. . JAMAICA.

GENUS PITANGUS.

CAUDIFASCIATUS. (D'ORB.)	CUBA and JAMAICA.
TAYLORI. SCL	. . PORTO RICO.
BAHAMENSIS. BRYANT	. . BAHAMAS.
GABBII. LAWR.	. SAN DOMINGO.

GENUS EMPIDONAX.

NANUS. LAWR.	SAN DOMINGO.
ACADICUS. (GMEL.)	. . CUBA.

GENUS CONTOPUS.

PALLIDUS. (GOSSE)	. . . JAMAICA.
LATIROSTRIS. VERR.	. . . ST. LUCIA.
BAHAMENSIS. (BRYANT)	. . . BAHAMAS.
HISPANIOLENSIS. (BRYANT)	. . SAN DOMINGO.
VIRENS. (LINN.).	. . . . CUBA.

GENUS SAYORNIS.

FUSCA. (GMEL.) .	. CUBA.

GENUS MYIARCHUS.

VALIDUS. CAB.	. JAMAICA.
STOLIDUS. (GOSSE)	. JAMAICA.
PHŒBE. (D'ORB.)	. CUBA and BAHAMAS.
ANTILLARUM. BRYANT	. PORTO RICO.
OBERI. LAWR.	ST. VINCENT, DOMINICA, GRENADA, ST. LUCIA.
SCLATERI. LAWR.	. MARTINIQUE.
DOMINICENSIS. (BRYANT)	. SAN DOMINGO.
CRINITUS. LINN.	. CUBA.

GENUS BLACICUS.

BARBIROSTRIS. (SW.)	. JAMAICA.
CARIBÆUS. (D'ORB.)	. CUBA.
BRUNNEICAPILLUS. LAWR.	. DOMINICA.
BLANCOI. GUNDL.	. PORTO RICO.

GENUS TYRANNUS.

ROSTRATUS. SCL.	. LESSER ANTILLES.
MAGNIROSTRIS. D'ORB.	CUBA and INAGUA.
MELANCHOLICUS. VIEILL.	. GRENADA.
DOMINICENSIS. (GMEL.) .	GREATER ANTILLES and ST. BARTHOLOMEW.
CAROLINENSIS. (GMEL.)	CUBA and PORTO RICO ?

FAMILY

COTINGIDÆ.

GENUS HADROSTOMUS.

NIGER. (GMEL.) .   . . .   . . JAMAICA.

FAMILY

CAPRIMULGIDÆ.

GENUS NYCTIBUS.

JAMAICENSIS. (GMEL.)	JAMAICA.
PALLIDUS. GOSSE	JAMAICA.

GENUS CHORDEILES.

MINOR. CABAN	ANTILLES.
VIRGINIANUS. (BRISS.)	CUBA and ANTILLES.

GENUS ANTROSTOMUS.

RUFUS. (BODD.)	ST. LUCIA.
CAROLINENSIS. (GMEL.)	BAHAMAS and GREATER ANTILLES.
CUBANENSIS. LAWR.	CUBA.

GENUS STENOPSIS.

CAYENNENSIS. (GMEL.) . . . . LESSER ANTILLES.

GENUS SIPHONORHIS.

AMERICANA. (LINN.) . . . . JAMAICA.

FAMILY
CYPSELIDÆ.

Genus CYPSELUS.

PHŒNICOBIUS. (Gosse) .   .   Jamaica, Cuba, and San Domingo.

Genus NEPHŒCETES.

NIGER. (Gmel.) .   .   .   .   .   Greater Antilles.

Genus CHÆTURA.

DOMINICANA. Lawr.   .   .   .   .   .   Dominica.

Genus HEMIPROCNE.

ZONARIS. (Shaw)   .   .   .   .   .   Jamaica and Cuba.

FAMILY
TROCHILIDÆ.

Genus GLAUCIS.

HIRSUTA. (Gmel.)   .   .   .   .   .   Grenada.

Genus LAMPORNIS.

DOMINICUS. (Linn.)   .   .   .   .   San Domingo and Porto Rico.
VIRIDIS. (Vieill.)   .   .   .   . Porto Rico.
MANGO. (Linn.) .   .   .   .   Jamaica.

Genus EULAMPIS.

JUGULARIS. (Linn.) .   .   .   .   .   . Lesser Antilles.
HOLOSERICEUS. (Linn.) .   .   .   .   . Lesser Antilles.

Genus AITHURUS.

POLYTMUS. (Linn.) .   .   .   .   Jamaica.

Genus THALURANIA.

BICOLOR. (Gmel.)   .   .   .   .   .   .   Dominica.

GENUS TROCHILUS.

COLUBRIS. LINN. . . . . CUBA and PORTO RICO.

GENUS MELLISUGA.

MINIMA. (LINN.) . . . . . . JAMAICA and HAITI.

GENUS CALYPTE.

HELENÆ. (LEMB.) . . . . CUBA.

GENUS DORICHA.

EVELYNÆ. (BOURC.). . . . . BAHAMAS.
LYRURA. GOULD. . . . . INAGUA and LONG ISLAND.

GENUS BELLONA.

CRISTATA. (LINN.) . ST. VINCENT, MARTINIQUE, BARBADOES, and ST. LUCIA.
EXILIS. (GMEL.). . . . DOMINICA, ST. THOMAS, NEVIS, and ST. CROIX.

GENUS SPORADINUS.

ELEGANS. (VIEILL.) . . . . SAN DOMINGO.
RICORDI. (GERB.) . . . . . . CUBA and BAHAMAS.
MAUGÆI. (VIEILL.) . . . . PORTO RICO.

FAMILY

TROGONIDÆ.

GENUS PRIONOTELES.

TEMNURUS. (TEMM.). . . . . CUBA.

GENUS TEMNOTROGON.

ROSEIGASTER. (VIEILL.) . . . . . . SAN DOMINGO.

FAMILY

CUCULLIDÆ.

GENUS CROTOPHAGA.

ANI. LINN. . . . . ANTILLES.

GENUS SAUROTHERA.

VETULA. (LINN.) . . . . JAMAICA.
DOMINICENSIS. LAFR. . . . . . . SAN DOMINGO.
VIEILLOTI. BP. . . . . PORTO RICO.
MERLINI. D'ORB. . . . . : . . CUBA.
BAHAMENSIS. BRYANT . . . . BAHAMAS.

GENUS COCCYGUS.

AMERICANUS. (LINN.)		BAHAMAS, CUBA, JAMAICA, PORTO RICO, and ST. CROIX.
MINOR. (GMEL.) .	. . . . . .	BAHAMAS and ANTILLES.
ERYTHROPHTHALMUS. (WILS.)	. . . . . .	. CUBA.

GENUS HYETORNIS.

PLUVIALIS. (GMEL.) .	. . . . .	JAMAICA.

FAMILY

ALCEDINIDÆ.

GENUS CERYLE.

TORQUATA. (LINN) .	. . .	. LESSER ANTILLES.
ALCYON. (LINN.) .	. . .	. . ANTILLES.

FAMILY

TODIDÆ.

GENUS TODUS.

VIRIDIS. LINN.		. . JAMAICA.
ANGUSTIROSTRIS. LAFR..		. SAN DOMINGO.
SUBULATUS. GOULD		. SAN DOMINGO.
PULCHERRIMUS. SHARPE		. . JAMAICA.?
HYPOCHONDRIACUS. BRYANT . . .		. . PORTO RICO.
MULTICOLOR. GOULD		. . . CUBA.

FAMILY

PICIDÆ.

GENUS PICUMNUS.

LAWRENCEI. CORY		HAITI and SAN DOMINGO.

GENUS CAMPEPHILUS.

PRINCIPALIS. (L.)
principalis bairdi. (*Cass.*) *Ridgw.*   . CUBA.

GENUS PICUS.

VILLOSUS. LINN.
villosus insularis. (*Maynard.*) *Cory* .   .   .   .   BAHAMAS.

Genus SPHYRAPICUS.
VARIUS. (Linn.) . . . . . Bahamas, Cuba, and Jamaica.

Genus XIPHIDIOPICUS.
PERCUSSUS. (Temm.) . . . . Cuba.

Genus MELANERPES.
PORTORICENSIS. (Daud.) . . . . . . Porto Rico.
L'HERMINIERI. (Less.) . . . . . . Guadeloupe.

Genus CENTURUS.
STRIATUS. (Mull.) . . . . San Domingo.
RADIOLATUS. (Wagl.) . . . . Jamaica.
SUPERCILIARIS. (Temm.) . . . . Cuba.

Genus COLAPTES.
CHRYSOCAULOSUS. Gundl. . . . . Cuba.

Genus NESOCELEUS.
FERNANDINÆ. (Vig.) . . . . Cuba.

FAMILY
PSITTACIDÆ.

Genus ARA.
TRICOLOR. (Bechst.) . . . . . Cuba and Jamaica.

Genus CONURUS.
EUOPS. (Wagl.) . . . . Cuba.
XANTHOLÆMUS. Sclater . . . . St. Thomas.
NANUS. (Vigors) . . . . Jamaica.
CHLOROPTERUS. (Souancé) . . . . . . San Domingo.
GUNDLACHI. Cabanis . . . . Mona.

Genus CHRYSOTIS.
SALLÆI. Scl. . . . . San Domingo.
VITTATA. (Bodd.) . . . . Porto Rico.
COLLARIA. (Linn.) . . . . Jamaica.
LEUCOCEPHALA. (Linn.) . . . . . . Bahamas and Cuba.
AGILIS. (Linn.) . . . . Jamaica.
AUGUSTA. (Vigors.) . . . . Dominica.
GUILDINGI. (Vigors.) . . . . St. Vincent.
BOUQUETI. (Bech.) . . . . Dominica.
VERSICOLOR. (Muller) . . . . St. Lucia.

FAMILY

STRIGIDÆ.

GENUS BUBO.

VIRGINIANUS. (GMEL.) (W. I. SCL. and SALV.) . JAMAICA?

GENUS STRIX.

FLAMMEA. (L.)

flammea fuscata. (*Temm.*) *Ridgw.*	. . . CUBA and JAMAICA.
flammea nigrescens. Lawr. . .	ST. VINCENT and DOMINICA.
flammea pratincola. Ridgw. .	. . . BAHAMAS.

GLAUCOPS. KAUP. SAN DOMINGO.

GENUS PSEUDOSCOPS.

GRAMMICUS. (GOSSE) . . . JAMAICA.

GENUS ASIO.

STYGIUS. (WAGL.) . . . . . CUBA.
ACCIPITRINUS. (PALL.) . . . . CUBA.
PORTORICENSIS. RIDGW. . . . PORTO RICO.

GENUS GYMNASIO.

NUDIPES. (DAUD.) . . . . . PORTO RICO, ST. JOHN, and ST. CROIX.
LAWRENCEII. (SCL. and SALV.) . . . . CUBA.

GENUS GLAUCIDIUM.

SIJU. (D'ORB.) . . CUBA.

GENUS SPEOTYTO.

CUNICULARIA. (MOLINA.)

cunicularia dominicensis. Cory . SAN DOMINGO and BAHAMAS.?

GUADELOUPENSIS. (RIDGW.) . . . . . GUADELOUPE and ST. NEVIS.
AMAURA. LAWR. . . . . ANTIGUA.

FAMILY

FALCONIDÆ.

Genus PANDION.

HALIÆTUS. (Linn.)

haliætus carolinensis. (*Gmel.*) (*Ridgw.*) .   Bahamas and Antilles.

Genus CIRCUS.

HUDSONICUS. (Linn.)   .   Cuba and Bahamas.

Genus RUPORNIS.

RIDGWAYI. Cory   San Domingo

Genus BUTEO.

BOREALIS. (Gmel.) .   .   . Bahamas, Porto Rico, Cuba, and Jamaica.
PENNSYLVANICUS. (Wils.)   .   .   Cuba, Porto Rico, and Antilles.

Genus ACCIPITER.

GUNDLACHI. Lawr. .   .   .   .   .   . Cuba.
FRINGILLOIDES. (Vig.) .   Cuba and San Domingo.
FUSCUS. (Gmel.)   .   .   .   .   . Bahamas.

Genus URUBITINGA.

ANTHRACINA. (Licht.) .   .   .   .   .   Cuba and St. Vincent.

Genus FALCO.

PEREGRINUS. Tunstall .   .   .   .   Bahamas and Antilles.
COLUMBARIUS. Linn.
              Cuba, Jamaica, San Domingo, Porto Rico, and St. Thomas.
SPARVERIUS. Linn. .   .   .   •   .   .   Bahamas and Antilles.

     sparverius dominicensis. (*Gmel.*) (*Ridgw.*) . Cuba? San Domingo, and Porto Rico.
     sparverius sparverioides. (*Vig.*) Cory   .   .   .   .   .   . Cuba.
     sparverius caribbæarum. (*Gmel.*) Cory   .   .   .   .   . Antilles

Genus ELANOIDES.

FORFICATUS. (Linn.)   .   .   Cuba and Jamaica.

Genus ROSTHRAMUS.

SOCIABILIS. (Vieill.)   .   .   .   .   •   .   .   . Cuba.

GENUS REGERHINUS.

WILSONII. (CASS.) . . . . CUBA.
*UNCINATUS. (TEMM.) . . . . . GRENADA.

GENUS POLYBORUS.

CHERIWAY. (JACQ.) . . . . CUBA.

FAMILY

CATHARTIDÆ.

GENUS CATHARTES.

AURA. (LINN.) . . . CUBA, JAMAICA, and BAHAMAS.

GENUS CATHARISTA.

ATRATA. (BARTR.) . . . . JAMAICA.

FAMILY

COLUMBIDÆ.

GENUS COLUMBA.

LEUCOCEPHALA. LINN. . . . BAHAMAS and ANTILLES.
CORENSIS. GMEL. . . . . ANTILLES.
CARIBÆA. LINN. . . . . JAMAICA and PORTO RICO.
INORNATA. VIGORS . . . . . GREATER ANTILLES.

GENUS ENGYPTILA.

JAMAICENSIS. (LINN.) . . JAMAICA.
WELLSI. LAWR. . . GRENADA.

GENUS ZENAIDURA.

CAROLINENSIS. (LINN.) . . . SAN DOMINGO, CUBA, and PORTO RICO.

---

* The note regarding the occurrence of this species in Grenada has been kindly forwarded to me by Mr. George N. Lawrence, specimens having been lately received by him from that island.

GENUS ECTOPISTES.

MIGRATORIA. (LINN)   .   .   . CUBA.

GENUS ZENAIDA.

AMABILIS. BP. . . . . ANTILLES.
MARTINICANA. BP. . . .   . .   . LESSER ANTILLES.
RUBRIPES. LAWR. . . .   . GRENADA.

GENUS MELOPELIA.

LEUCOPTERA. (LINN.)   . . . SAN DOMINGO, JAMAICA, and CUBA.

GENUS CHAMÆPELIA.

PASSERINA. (LINN.) .   BAHAMAS and ANTILLES.

GENUS GEOTRYGON.

CRISTATA. (TEMM.) . . .   . JAMAICA.
MYSTACEA. (TEMM.) . . .   . . GUADELOUPE and ST. LUCIA.
CANICEPS. GUNDL. .   .   . . CUBA.
MONTANA. (LINN.) .   .   . . . . ANTILLES.
MARTINICA. (GMEL.)   .   . . BAHAMAS and ANTILLES.

GENUS STARNŒNAS.

CYANOCEPHALA. (LINN.)   .   . . . . CUBA.

FAMILY

PHASIANIDÆ.

GENUS NUMIDA.

MELEAGRIS. LINN. (Introduced) . . .   . ANTILLES.

FAMILY

TETRAONIDÆ.

GENUS COLINUS.

CUBANENSIS. (GOULD) .   .   .   .   . CUBA and PORTO RICO.
VIRGINIANUS. (LINN.)
        BAHAMAS, SAN DOMINGO, PORTO RICO, JAMAICA, ST. CROIX, and ANTIGUA.

GENUS EUPSYCHORTYX.

SONNINII. (TEMM.) . . . . ST. THOMAS.

FAMILY
ŒDICNEMIDÆ.

Genus ŒDICNEMUS.
DOMINICENSIS. Cory . . . . . San Domingo.

FAMILY
CHARADRIDÆ.

Genus SQUATAROLA.
HELVETICA. (Linn.). . . . . . . Bahamas and Antilles.

Genus CHARADRIUS.
DOMINICUS. Müll. . . . . Bahamas and Antilles.

Genus ÆGIALITIS.
VOCIFERUS. (Linn.) . . . . Bahamas and Greater Antilles.
WILSONIUS. (Ord.) . . . . .Bahamas and Greater Antilles.
SEMIPALMATUS. (Bp.) . . . . . Bahamas and Antilles.
NIVOSUS. Cass. . . . . Cuba.
MELODUS. (Ord.) . . . . .Bahamas and Greater Antilles.

FAMILY
HÆMATOPIDÆ.

Genus HÆMATOPUS.
PALLIATUS. (Temm.) . . . Bahamas and Greater Antilles.

FAMILY
STREPSILIDÆ.

Genus STREPSILAS.
INTERPRES. (Linn.) . . . . . . Bahamas and Antilles.

FAMILY
RECURVIROSTRIDÆ.

Genus HIMANTOPUS.

NIGRICOLLIS. (Vieill.) .   .   Bahamas and Antilles.

Genus RECURVIROSTRA.

AMERICANA. (Gmel.)   .   .   .   .   Cuba and Jamaica.

FAMILY
SCOLOPACIDÆ.

Genus GALLINAGO.

WILSONI. (Temm.)   .   .   .   .   Bahamas and Antilles.

Genus MACRORHAMPHUS.

GRISEUS. (Gmel.)   .   .   .   .   .   Cuba, Jamaica, and Bahamas.
SCOLOPACEUS. (Say.)   .   .   .   .   .   .   Cuba and Antilles.

Genus MICROPALAMA.

HIMANTOPUS. (Bp.)   .   .   .   .   Antilles.

Genus EREUNETES.

PUSILLUS. (Linn.) .   .   .   .   Antilles.
OCCIDENTALIS. Lawr.   .   .   .   .   .   Antilles.

Genus TRINGA.

MINUTILLA. (Vieill.)   .   .   .   .   Antilles.
MACULATA. (Vieill.)   .   .   .   .   Antilles.
FUSCICOLLIS. Vieill.   .   .   .   .   Antilles.
CANUTUS   .   .   .   .   Jamaica?

Genus CALIDRIS.

ARENARIA. (Linn.) .   .   .   .   .   Antilles.

Genus LIMOSA.

FEDOA. (Linn.) .   .   .   .   .   Greater Antilles.
HÆMASTICA. (Linn.)   .   .   .   .   .   .   Cuba.

GENUS SYMPHEMIA.
SEMIPALMATA. (GMEL.) . . . . . BAHAMAS and ANTILLES.

GENUS TOTANUS.
MELANOLEUCUS. GMEL. . . . . . . BAHAMAS and ANTILLES.
FLAVIPES. GMEL. . . . . ANTILLES.

GENUS RHYACOPHILUS.
SOLITARIUS. (WILS.) . . . . ANTILLES.

GENUS TRINGOIDES.
MACULARIUS. (LINN.) . . . . . ANTILLES.

GENUS BARTRAMIA.
LONGICAUDA. (BECHST.) . . . . . BAHAMAS, CUBA, and JAMAICA.

GENUS TRINGITES.
RUFESCENS. (VIEILL.) . . . . . CUBA.

GENUS NUMENIUS.
HUDSONICUS. (LATH.) . . . . ANTILLES.
BOREALIS. (FORST.) . . . . ANTILLES.
LONGIROSTRIS. (WILS.) . . . . GREATER ANTILLES.

FAMILY

CICONIIDÆ.

GENUS TANTALUS.
LOCULATOR. LINN. . . . . CUBA.

FAMILY

IBIDIDÆ.

GENUS EUDOCIMUS.
ALBUS. (LINN.) . . . . . . GREATER ANTILLES.
RUBER. (LINN.) . . . . . JAMAICA and CUBA

GENUS PLEGADIS.
FALCINELLUS. (LINN.) . . . . . GREATER ANTILLES.

28

FAMILY

PLATALEIDÆ.

Genus AJAJA.

ROSEA. (Linn.) . . Bahamas, Greater Antilles, and Grenada.

FAMILY

PHŒNICOPTERIDÆ.

Genus PHŒNICOPTERUS.

RUBER. (Linn.) . . . . . Bahamas and Greater Antilles.

FAMILY

ARDEIDÆ.

Genus ARDEA.

HERODIAS. (Linn.) . . . . Bahamas and Antilles.
EGRETTA. (Gmel.) . . . Bahamas, Greater Antilles, and Barbuda.
CANDIDISSIMA. (Gmel.) . . . . . Bahamas and Antilles.
RUFA. (Bodd.) . . . . . Bahamas and Greater Antilles.
CÆRULEA. (Linn.) . . . . . . Bahamas and Antilles.
VIRESCENS. (Linn.) . . . . . . Bahamas and Antilles.
OCCIDENTALIS. Aud. . . . Porto Rico, Jamaica, and Cuba.
BRUNNESCENS. Gundl. . . . . Cuba.
TRICOLOR. (Müll.) . . . . Bahamas and Greater Antilles.
? CYANIROSTRIS. Cory . . . . . . Inagua and Bahamas. ?

Genus NYCTIARDEA.

VIOLACEA. (Linn.) . . . . . Cuba and Bahamas.
GRISEA. (Bodd.)
 grisea nævia. (Briss) Allen . . . . Greater Antilles.

Genus BOTAURUS.

LENTIGINOSUS. (Mont.). . . . . Jamaica, Cuba, and Porto Rico.

Genus ARDETTA.

EXILIS. (Gmel.). . . . . Greater Antilles.

29

FAMILY
GRUIDÆ.

Genus GRUS.

MEXICANA. Müll. . . . . Cuba.

FAMILY
ARAMIDÆ.

Genus ARAMUS.

GIGANTEUS. (Bonap.) . . . Greater Antilles.

FAMILY
PARRIDÆ.

Genus PARRA.

VIOLACEA. Cory . Cuba and San Domingo.

FAMILY
RALLIDÆ.

Genus RALLUS.

MACULATUS. (Bodd.) . . . . Cuba.
ELEGANS. Aud. . . . . Cuba.
VIRGINIANUS. Linn. . . . . Cuba.
LONGIROSTRIS. (Bodd.)
  longirostris crepitans. (*Gmel.*) *Ridgw.* . . . . . . Bahamas.
  longirostris caribæus. *Ridgw.* . . . . Antilles.

Genus PORZANA.

CONCOLOR. (Gosse) . . . . Jamaica.
FLAVIVENTRIS. (Bodd.) . . . . . Jamaica and Cuba.
JAMAICENSIS. (Gmel.) . . . . . Jamaica and Cuba.
CAROLINA. (Linn.) . . . . . Bahamas and Antilles.
NOVEBORACENSIS. (Gmel.) . . . . Cuba.

Genus GALLINULA.

GALEATA. (Licht.) . . . Bahamas and Antilles.

Genus IONORNIS.

MARTINICA. (Linn.) . Bahamas and Antilles

Genus FULICA.

AMERICANA. Gmel. . . . . . Bahamas and Antilles.
CARIBÆA. Ridgw. . . . . Antilles.

FAMILY
ANATIDÆ.

Genus ANSER.

GAMBELII. Hartl. . . . . Cuba.

Genus CHEN.

HYPERBOREUS. (Pall.) . . . . . Cuba, Porto Rico, and Jamaica.
CÆRULESCENS. (Linn.) . . . . . . Bahamas and Cuba.

Genus BERNICLA.

CANADENSIS. (Linn.) . . . . Jamaica?

Genus DENDROCYGNA.

ARBOREA. (Linn.) . . . . . . Bahamas and Antilles.
AUTUMNALIS. (Linn.) . . . . Jamaica.
VIDUATA. (Linn.) (Introduced) . . . . Cuba.

Genus ANAS.

STREPERUS. (Linn.) . . . . . . Cuba and Jamaica.
BOSCHAS. Linn. . . . . . Cuba, Bahamas, and Jamaica.
VALLISNERIA. Wils. . . . . . . Cuba and Jamaica.
OBSCURA. Gmel. . . . . Cuba and Jamaica.

GENUS DAFILA.

BAHAMENSIS. (LINN.) . . . . . BAHAMAS and ANTILLES.
ACUTA. (LINN.) . . . . . . JAMAICA, PORTO RICO, and CUBA.

GENUS MARECA.

AMERICANA. (GMEL.) . . . . PORTO RICO, CUBA, and JAMAICA.

GENUS QUERQUEDULA.

DISCORS. (LINN.) . . . . BAHAMAS and ANTILLES.
CAROLINENSIS. (GMEL.) . . . . . CUBA, JAMAICA, and BAHAMAS.

GENUS SPATULA.

CLYPEATA. (LINN.) . . . . . PORTO RICO, CUBA, and JAMAICA.

GENUS AIX.

SPONSA. (LINN.) . . . CUBA and JAMAICA.

GENUS FULIGULA.

AFFINIS. EYTON . . . . . PORTO RICO, CUBA, BAHAMAS, and JAMAICA.
COLLARIS. (DONOV.) . . . PORTO RICO, CUBA, BAHAMAS, and JAMAICA.
AMERICANA. EYTON . . . . . CUBA, BAHAMAS, and JAMAICA.

GENUS CLANGULA.

ALBEOLA. (LINN.) . . . . CUBA.
GLAUCIUM. (LINN.) . . . . . CUBA and BARBUDA.

GENUS ŒDEMIA.

PERSPICILLATA. (LINN.) . . . . . JAMAICA.?

GENUS ERISMATURA.

RUBIDA. (WILS.) . . BAHAMAS, JAMAICA, PORTO RICO, and CUBA.

GENUS NOMONYX.

DOMINICUS. (LINN.) . . . . . ANTILLES.

GENUS MERGUS.

CUCULLATUS. LINN. . . . . . CUBA and ANTILLES.?

FAMILY

TACHYPETIDÆ.

GENUS FREGATA.

AQUILA. (LINN.) . . . . BAHAMAS and ANTILLES.

FAMILY

PELECANIDÆ.

Genus PELECANUS.

FUSCUS. (Linn.)   .   .   .   Bahamas and Antilles.

FAMILY

PHALACROCORACIDÆ.

Genus PHALACROCORAX.

DILOPHUS. (Sw.)

dilophus floridanus. (*Aud.*) *Ridgw.* .   Bahamas and Cuba.

MEXICANUS. (Brandt)   .   .   .   . Cuba.

FAMILY

PLOTIDÆ.

Genus PLOTUS.

ANHINGA. Linn.   .   .   .   . Cuba.

FAMILY

SULIDÆ.

Genus SULA.

CYANOPS. (Sundev.) .   .   .   .   .   . Bahamas and Antilles.

LEUCOGASTRA. (Bodd.) .   .   .   . Antilles.

PISCATOR. (Linn.)   .   .   .   .   .   . Cuba, Jamaica, and Antilles

33

FAMILY

PHÆTHONTIDÆ.

Genus PHÆTHON.

FLAVIROSTRIS. Brandt . . . . . . Bahamas and Antilles.
ÆTHEREUS. (Linn.) Guadeloupe, St. Vincent, Grenada, Cuba, and Jamaica.

FAMILY

RHYNCHOPIDÆ.

Genus RHYNCHOPS.

NIGRA. Linn. . . . . . . Cuba and St. Croix?

FAMILY

LARIDÆ.

Genus LARUS.

ATRICILLA. Linn. . . . . . . Bahamas and Greater Antilles.
ARGENTATUS. Brunn. . . . . Bahamas and Cuba.
DELAWARENSIS. Ord. . . . . Cuba. ?
? FRANKLINI. Sw. . . . . Antilles. ?

Genus STERNA.

ANOSTHÆTA. Scop. . . . . Bahamas and Antilles.
ANGLICA. Mont. . . . . Bahamas and Antilles.
CANTIACA. (Gmel.) . . . . Bahamas and Antilles.
HIRUNDO. Linn. . . . . Bahamas.
FULIGINOSA. Gmel. . . . . . Bahamas and Antilles.
MAXIMA. Bodd. . . . . . . Bahamas and Antilles.
DOUGALLI. Mont. . . . . . Bahamas and Antilles.
ANTILLARUM. (Less.) . . . . Bahamas and Antilles.

34

GENUS HYDROCHELIDON.

LARIFORMIS. (LINN.)   .   ANTILLES.

GENUS ANOUS.

STOLIDUS. (LINN.) .   BAHAMAS and ANTILLES.

FAMILY

PROCELLARIIDÆ.

GENUS OCEANITES

OCEANICUS. (WILS.)   .   BAHAMAS and GREATER ANTILLES.

GENUS ŒSTRELATA.

HESITATA. (KUHL.) .   .   .   .   .   ANTILLES.
JAMAICENSIS. (BANCR.) .   .   .   .   JAMAICA.

GENUS PUFFINUS.

MAJOR. FABER .   .   .   .   BAHAMAS.
AUDUBONI. FINSCH .   .   .   .   .   BAHAMAS and ANTILLES.

FAMILY

PODICIPIDÆ.

GENUS PODICEPS.

DOMINICUS. (LINN.)   .   .   BAHAMAS and GREATER ANTILLES.

GENUS PODILYMBUS.

PODICEPS. (LINN.) .   .   .   .   ANTILLES.

APPENDIX.

The following species have been recorded from the West Indies : —

AMMADROMUS MARITIMUS. (Wils.) .   .   . Cuba.
    Caban, J. F. O., p. 7, 1856.

CHRYSOMITRIS MEXICANA. Swain.   .   . .   , Cuba.
    Gundl. Rep. Fis. Nat. Cuba, i, p. 397, 1866.

CHRYSOMITRIS PINUS. (Bartr.) .   .   .   . Cuba.
    Gundl. J. F. O., p. 9, 1856, and Rep. Fis. Nat. Cuba, p. 397, 1866.

CARDUELES ELEGANS. Steph.   .   . Cuba.
    Probably an escaped cage bird.

CARDINALIS VIRGINIANUS. Bp. .   .   .   . . Cuba.
    Gundl. Rep. Fis. Nat. Cuba, p. 397, 1866.

MOLOTHRUS BONARIENSIS. (Cab.)   .   .   . St. Thomas.
    M. Sericeus, Newton Ibis, p. 308, 1860.

CYANEOCORAX PILEATUS. (Zemm.)   Jamaica.
    Probably an escaped cage bird.

ARA MILITARIS. (Linn.) .   .   Jamaica and Cuba.
    Probably A. tricolor.

RUPORNIS MAGNIROSTRIS. (Gm.)   Martinique.
    Gurney, Ibis, p. 482, 1876.

TURTER RISORIA. (Linn.)   .   Jamaica, Cuba, and St. Bartholomew.
    Introduced species.

ORTALIDA RUFICAUDA. Jard.   .   .   . Union I. Grenadines.
    Introduced species.

ANAS MAXIMA. Gosse.   .   .   Jamaica.
    Supposed to be a hybrid.

CAIRINA MOSCHATA. (Linn.)   Jamaica, Cuba.
    Introduced species.

QUERQUEDULA CYANOPTERA. (Vieill.)   .   .   . Cuba.
    Brewer, Proc. Bost. Soc. Nat. Hist., p. 308, 1860.

NYROCA FERRUGINEA. (Gm.)   .   .   .   .   Jamaica.
    Probably some other species wrongly identified.

LARUS PHILADELPHIA. (Ord.)   .   .   .   . Bahamas.
    Claimed to have been observed by Moore.

www.ingramcontent.com/pod-product-compliance
Lightning Source LLC
Chambersburg PA
CBHW022005190326
41519CB00010B/1394